551.6   Witty, Margot.
WIT
         A day in the life of
         a meteorologist.

                          33197000037668

$11.79

DATE			

A DAY IN THE LIFE OF A
Meteorologist

by Margot and Ken Witty
Photography by Stephen Sanacore

Troll Associates

Library of Congress Catalog Card Number: 80-54098
ISBN 0-89375-450-1   ISBN 0-89375-451-X Paper Edition

The authors and publisher wish to thank Joe Witte and his associates at KYW-TV 3 in Philadelphia for their
generous cooperation and assistance.

Photograph credits: pp. 9, 15—National Oceanic and Atmospheric Administration; p. 13—National Aeronautics
and Space Administration; pp. 6, 14, 15, 17, 26, 28—Joe Witte.

When Joe Witte goes for his morning run one day in
late November, the sky is bright and clear. But Joe is
expecting trouble. He is a meteorologist, and his job
is predicting the weather. For the last two days he
has been tracking a major storm that is moving
toward his area.

Normally, Joe starts work sometime after lunch, and works until almost midnight. But today, because he knows the storm is only hours away, he gets an early start. The sky is still bright as he walks to the office just before nine.

Joe works for a television station. He usually does
two broadcasts a day, one in the early evening and
one on the late news. Each day, he tries to forecast
the weather for the next five days.

At the office, the first thing Joe does is check the radar monitor. Meteorologists use radar to look for clouds, rain, or snow. Weather radar uses echoes from radio waves to detect moisture. The radio waves are sent out by transmitters that are located at airports.

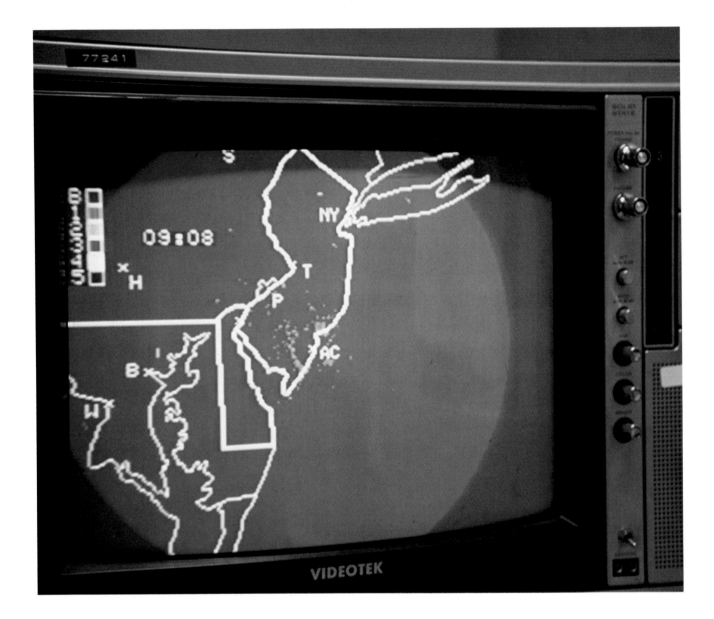

If the radio waves strike drops of moisture, their echoes show up as dots of colored light on the monitor screen. Although Joe sees the dots that mean light rain nearby, the storm has not yet reached his local area.

Next, Joe checks National Weather Service maps of the U.S. and Canada. These maps show how weather systems are expected to move across the country during the next twelve hours. Joe uses them to help him predict what will happen in his own area.

The National Weather Service collects weather information from airports, ships, and high-altitude balloons. The information includes moisture, temperature, air pressure, and wind measurements from thousands of locations.

Joe now begins to gather information about his local area. The first step is to check the anemometer, or wind-speed indicator, on the roof. Rising winds warn him that the storm is getting closer. Then he plugs in by telephone to the National Weather Service computer center.

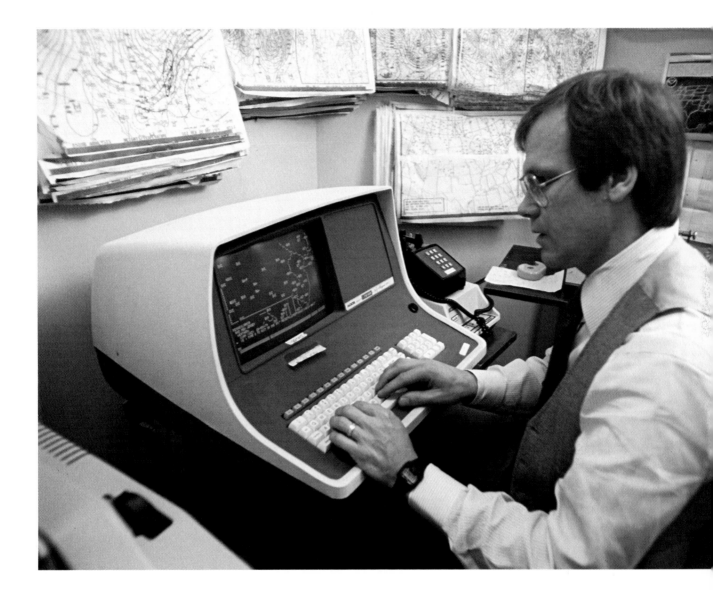

By typing questions into his computer terminal, Joe can get up-to-the-minute facts from any airport in the U.S. or Canada. He asks for the high- and low-temperature readings recorded at a nearby airport. Local temperature is rising—another sign of the approaching storm.

A high-speed printer records the computer information on a map. Joe marks the progress of the storm on the print-out map. He draws a blue line between a warm air mass and the cold front that marks the leading edge of the rain.

Joe also studies the recent photographs of the U.S. taken by a weather satellite orbiting the earth. Each map shows that the storm system is getting nearer the city. By late morning, clouds have covered the sky, and rain is beginning to fall.

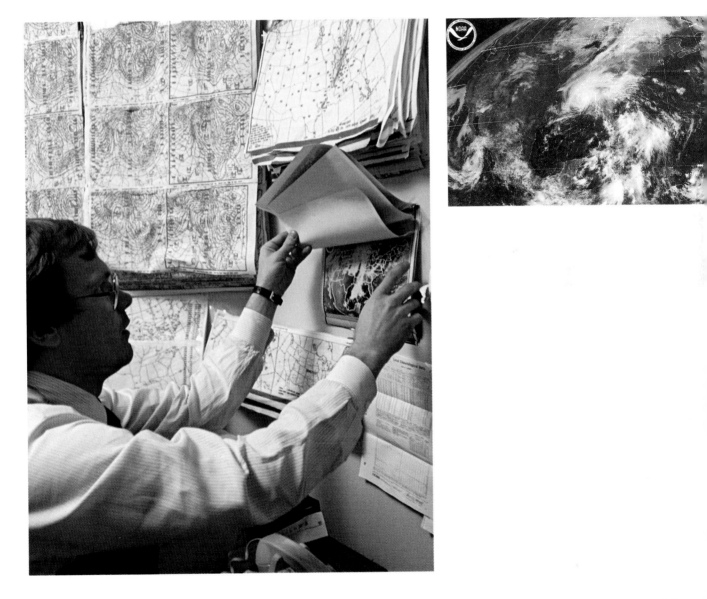

Will the center of the storm hit the city? It is still too soon to be sure. Joe decides to go out and pick up a sandwich for lunch. His equipment can tell him a lot about the weather, but getting outside—getting a feel for the day—makes him more certain of his local forecast.

Whenever he is outside in bad weather, Joe is re-
minded of all the people who depend on his fore-
casts. It's not just the farmers and fishermen who
are at the mercy of the weather. For city dwellers, a
bad storm can mean snarled or slow-moving traffic,
automobile accidents, and power failures.

Back at his desk, Joe eats his lunch, and talks to a friend who is a meteorologist in a neighboring state. They compare notes on today's storm. There has been a lot of rain already this fall. More rain from this storm could cause flash floods throughout the region.

Outside, the rain is falling harder. Joe checks a teletype machine that provides continuous reports from nearby weather stations. A special bulletin has just been issued, warning of flash floods in the central part of the state. The National Weather Service asks that the bulletin be broadcast immediately.

In his desk, Joe keeps a special set of slides for times like these. He pulls out the one that says, "Flash Flood Watch," and hurries to the control room. There he tells a technician that he will be issuing a weather bulletin in the next few minutes. "Have the slide ready to put on the air," he says.

A program is interrupted so Joe can go on the air.
He says, "The National Weather Service has issued
a flash-flood watch, effective until 9:30 tonight, for
the central portion of the state. People in low-lying
areas should leave their homes and move to higher
ground."

Joe goes to the daily planning meeting for the evening news programs. He is asked how serious the storm is likely to be. "I've just put out a flash flood watch for the center of the state," says Joe. "But there's still a chance the worst of the storm will miss the city."

Whatever happens, they decide to give the weather story a top spot on the evening news. Joe and a videotape technician look at a satellite film from the National Weather Service. The film will show TV viewers how the storm has developed.

While Joe is watching the film, his young friend Billy comes to visit. The two became friends after Billy invited Joe to give a talk to his junior-high-school science class. Joe shows Billy a videotape from last year, when he covered a big hurricane. The hurricane caused widespread damage.

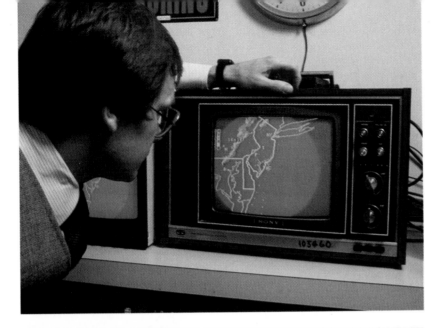

But there is not much time to talk today. On the radar screen, Joe can see the dots that indicate the location of heavy rain. He asks Billy to collect the latest satellite photos. They show that thick clouds now cover most of the East Coast.

It is getting close to air time. Joe must write his weather report. He uses all the information he has collected during the day to decide what to tell his viewers. Then Joe dabs makeup on his face to keep his skin from shining under the hot lights used during the broadcast.

In the studio, Joe gives the figures that will accompany his report to a stagehand. The stagehand arranges the numbers on a map and on the large board that shows Joe's forecast for the next five days.

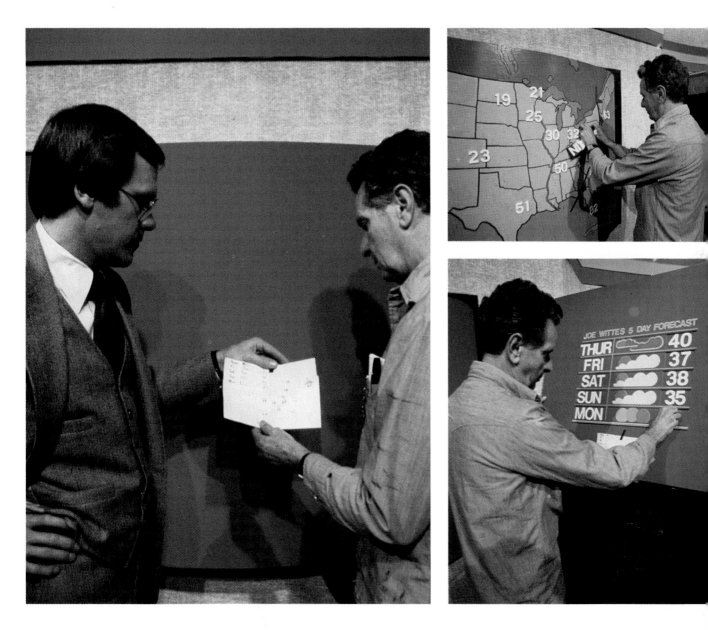

At 5:37, Joe goes on the air. "Heavy rains fell on the center of the state today, causing floods in many areas. Scenes like this one were the result. Many people had to be evacuated. Here in the city, the rain is lighter, but there has been some isolated flooding. Rush-hour driving may be hazardous, so drive carefully."

Then Joe gives his forecast. He predicts that the storm system that brought the rain to his viewing area will move north, bringing snow to the Great Lakes. "But that's good news for us," he adds, "because it should leave us with much drier air for the next few days."

Later, as Joe and Billy tour the control room, they talk about what it takes to become a meteorologist. Joe's training included a lot of math and physics. But his special interest was glaciers. As a graduate student, he lived on an ice island in Alaska, and climbed the South Cascade Glacier in Washington State.

Joe's apartment is a few minutes' walk from the station, so Joe can go home for dinner. Over supper, he tells his wife about Billy's visit. Joe would like to get more boys and girls interested in meteorology, perhaps through a TV science series for young people.

Joe goes back to work every night after supper, to prepare his late report. By now the storm has passed over the city, though rain is still falling in some areas. Joe marks the storm's progress on his map, and then contacts a suburban airport to check on conditions there.

Shortly before 11:30, while TV viewers are relaxing at home, Joe gives his last broadcast of the day. "This city suffered little damage from one of the fall's worst storms. But severe flooding has paralyzed the state capital, and trees are down along many highways. Fortunately, no lives were lost."

When Joe Witte came to work, a major storm was threatening the city. Now, almost fifteen hours later, the storm has passed, and Joe can finally call it a day. With winter and its many snowstorms ahead, long days like this one won't be unusual. But they're all part of the job of a meteorologist.